AUG -- 2021

Jefferson Township Public Library
1031 Weldon Road
Oak Ridge, N.J.
phone: 973-208-6244
http://www.jeffersonlibrary.net

Fall Holidays

Julie Murray

Abdo Kids Junior
is an Imprint of Abdo Kids
abdobooks.com

SEASONS: FALL FUN!

abdobooks.com

Published by Abdo Kids, a division of ABDO, P.O. Box 398166, Minneapolis, Minnesota 55439.
Copyright © 2021 by Abdo Consulting Group, Inc. International copyrights reserved in all countries.
No part of this book may be reproduced in any form without written permission from the publisher.
Abdo Kids Junior™ is a trademark and logo of Abdo Kids.

Printed in the United States of America, North Mankato, Minnesota.

052020

092020

Photo Credits: iStock, Shutterstock

Production Contributors: Teddy Borth, Jennie Forsberg, Grace Hansen

Design Contributors: Candice Keimig, Pakou Moua, Dorothy Toth

Library of Congress Control Number: 2019955589
Publisher's Cataloging-in-Publication Data

Names: Murray, Julie, author.
Title: Fall holidays / by Julie Murray
Description: Minneapolis, Minnesota : Abdo Kids, 2021 | Series: Seasons: fall fun! | Includes online resources and index.
Identifiers: ISBN 9781098202170 (lib. bdg.) | ISBN 9781098203153 (ebook) | ISBN 9781098203641 (Read-to-Me ebook)
Subjects: LCSH: Autumn--Juvenile literature. | Holidays--Juvenile literature. | Seasons--Juvenile literature.
Classification: DDC 525.5--dc23

Table of Contents

Fall Holidays 4

More Fall
Fun Days 22

Glossary 23

Index 24

Abdo Kids Code 24

Fall Holidays

There are many fall holidays.

It is Labor Day! It's time for a picnic!

Ruth blows a horn.

It is Rosh Hashanah!

Rani lights an oil lamp.

It is Diwali!

Mia gets her face painted.

It is Halloween!

Rosa is in a parade. It is Day of the Dead!

Caleb holds a flag. It is Veterans Day!

Jamie eats turkey. It is Thanksgiving!

What fall holidays do you celebrate?

More Fall Fun Days

Read a Book Day
(September 6)

Taco Day
(October 4)

World Kindness Day
(November 13)

Take a Hike Day
(November 17)

Glossary

Day of the Dead
a yearly Mexican celebration to honor the spirits of the dead, observed on November 1 and 2.

Diwali
a Hindu festival of lights, held between October and November.

Labor Day
a public holiday held in honor of working people in September in North America.

Rosh Hashanah
a Jewish New Year festival that is held in September.

Index

Day of the Dead 14

Diwali 10

Halloween 12

Labor Day 6

Rosh Hashanah 8

Thanksgiving 18

Veterans Day 16

Visit **abdokids.com** to access crafts, games, videos, and more!

Use Abdo Kids code

SFK2170

or scan this QR code!